Number Sense Bingo Book

COMPLETE BINGO GAME IN A BOOK

Written By Rebecca Stark
Educational Books 'n' Bingo

TITLE: Number Sense Bingo
AUTHOR: Rebecca Stark

ISBN 978-0-87386-458-9

Educational Books 'n' Bingo
Printed in the U.S.A.

NUMBER SENSE BINGO DIRECTIONS

INCLUDED:

List of Terms

Templates for Additional Terms and Clues

2 Clues per Term

30 Unique Bingo Cards

Markers

1. **Either cut apart the book or make copies of ALL the sheets. You might want to make an extra copy of the clue sheets to use for introduction and review. Keep the sheets in an envelope for easy reuse.**

2. Cut apart the call cards with terms and clues.

3. Pass out one bingo card per student. There are enough for a class of 30.

4. Pass out markers. You may cut apart the markers included in this book or use any other small items of your choice.

5. Decide whether or not you will require the entire card to be filled. Requiring the entire card to be filled provides a better review. However, if you have a short time to fill, you may prefer to have them do the just the border or some other format. Tell the class before you begin what is required.

6. There are 50 terms. Read the list before you begin. If there are any terms that have not been covered in class, you may want to read to the students the term and clues before you begin.

7. There is a blank space in the middle of each card. You can instruct the students to use it as a free space or you can write in answers to cover terms not included. Of course, in this case you would create your own clues. (Templates provided.)

8. Shuffle the cards and place them in a pile. Two or three clues are provided for each term. If you plan to play the game with the same group more than once, you might want to choose a different clue for each game. If not, you may choose to use more than one clue.

9. Be sure to keep the cards you have used for the present game in a separate pile. When a student calls, "Bingo," he or she will have to verify that the correct answers are on his or her card AND that the markers were placed in response to the proper questions. Pull out the cards that are on the student's card keeping them in the order they were used in the game. Read each clue as it was given and ask the student to identify the correct answer from his or her card.

10. If the student has the correct answers on the card AND has shown that they were marked in response to the *correct questions,* then that student is the winner and the game is over. If the student does not have the correct answers on the card OR he or she marked the answers in response to *the wrong questions,* then the game continues until there is a proper winner.

11. If you want to play again, reshuffle the cards and begin again.

Have fun!

TERMS/ANSWERS

>	ESTIMATE
<	EXPONENT
-3	FACTORS
1/2	HUNDREDTHS PLACE
7/24	IMPROPER FRACTION
11/20	INTEGERS
4/7	INVERSE OPERATIONS
4/5	IRRATIONAL NUMBER
.875	MIXED NUMBER
1 3/4	MULTIPLE
2 2/3	NEGATIVE INTEGER
6 2/3	PERCENT
7	PLACE VALUE
9	POSITIVE INTEGER
15%	PRIME FACTORIZATION
36	PRIME NUMBER
42	PROPER FRACTION
48.307	RATIONAL NUMBERS
400%	RECIPROCAL
ABSOLUTE VALUE	REDUCE
ASSOCIATIVE PROPERTY	ROUND
COMMUTATIVE PROPERTY	THOUSANDS PLACE
COMPOSITE NUMBER	THOUSANDTHS PLACE
DECIMAL	WHOLE NUMBERS
DISTRIBUTIVE PROPERTY	ZERO

Additional Terms

Choose as many additional terms as you would like and write them in the squares. Repeat each as desired.
Cut out the squares and randomly distribute them to the class.
Instruct the students to place their square on the center space of their card.

Number Sense Bingo

Clues for Additional Terms

Write three clues for each of your terms.

_____	_____
1.	1.
2.	2.
3.	3.
_____	_____
1.	1.
2.	2.
3.	3.
_____	_____
1.	1.
2.	2.
3.	3.

= ≠	= ≠	= ≠	= ≠	= ≠
= ≠	= ≠	= ≠	= ≠	= ≠
= ≠	= ≠	= ≠	= ≠	= ≠
= ≠	= ≠	= ≠	= ≠	= ≠
= ≠	= ≠	= ≠	= ≠	= ≠
= ≠	= ≠	= ≠	= ≠	= ≠
= ≠	= ≠	= ≠	= ≠	= ≠

>
1. This sign means "greater than."
2. Which sign should be inserted?
     3 ___ 2; 100 ___ 99; 5 ___ 1.
3. Which sign should be inserted?
    200 ___ 199; 99 ___ 97; 23 ___ 21.
	<
	1. This sign means "less than."
	2. Which sign should be inserted?
	    2 ___ 3; 99 ___ 100; 1 ___ 5.
	3. Which sign should be inserted?
	    199 ___ 200; 97 ___ 99; 21 ___ 23.

-3 1. Its absolute value is 3. 2. -281 + 278 = 3. -27 ÷ 9 =	**1/2** 1. .5 2. 50% 3. 50/100
7/24 1. 7/6 x 1/4 = 2. 7/6 ÷ 4 = 3. 5/12 − 1/8 =	**11/20** 1. 3/4 − 1/5 = 2. 1/4 = 3/10 = 3. 1 − 9/20 =
4/7 1. It is 20/35 reduced to its lowest terms. 2. The reciprocal of 7/4. 3. 1 2/7− 5/7 =	**4/5** 1. 80% 2. Reciprocal of 5/4. 3. .80
.875 1. 7/8 2. 14/16 3. .175 x 5 =	**1 3/4** 1. 1/2 ÷ 2/7 = 2. 1.75 3. 7/4

Number Sense Bingo

2 2/3	6 2/3
1. 2/3 ÷ 1/4 = 2. 2/3 x 4 = 3. 12 − 9 1/3 =	1. 5/6 ÷ 1/8 = 2. 5/6 x 8 = 3. 9 11/12 − 3 1/4 =

7	9
1. The greatest common factor (GCF) of 14 and 19. 2. $\sqrt{49}$ 3. 700/100	1. 1.08 ÷ .12 = 2. The lowest common multiple (LCM) of 3 and 9. 3. The greatest common factor (GCF) of 27 and 18.

15%	36
1. .15 2. 15/100 3. 30 is ___ of 200.	1. Absolute value of -36. 2. 6^2 3. 14.27 + 21.73 =

42	48.307
1. The lowest common multiple (LCM) of 6 and 7. 3. 10% of 420. 3. 20% of 210.	1. 7.21 x 6.7 = 2. 69.495 − 21.188 = 3. 40.007 + 8.3 =

400%	ABSOLUTE VALUE
1. 4.0 2. 400/100 3. 16 is ___ of 4.	1. How far an integer is from zero. 2. Put the integer between two parallel vertical lines to show this. 3. The ___ of -2 is 2. It is written I-2I = 2.

Number Sense Bingo

ASSOCIATIVE PROPERTY 1. This might be called the Grouping Property. 2. Refers to the fact that changing the grouping the addends does not change the sum. 3. Refers to the fact that changing the grouping of factors does not change the product.	**COMMUTATIVE PROPERTY** 1. It might also be called the Order Property. 2. Refers to the fact that the order of addends does not affect the sum. 3. The fact that the order of the factors does not affect the product.
COMPOSITE NUMBER 1. A whole number with at least one factor besides itself and 1. 2. Two is the only even number that is not this kind of number. 3. A number divisible by at least one number other than itself and 1.	**DECIMAL** 1. In this kind of number the digits to the right of the point have a value less than 1. 2. To change a fraction to one, divide the numerator by the denominator. 3. To convert the ___ .045 into a fraction, first change it to 45/1000 and then reduce the fraction to 9/200.
DISTRIBUTIVE PROPERTY 1. Refers to the fact that the product of the sum or difference of 2 or more numbers is the same as the sum or difference of their products. 2. The following is an example of ___: a (b + c) = ab + ac. 3. The following is an example of ___: 5 x (3 + 4) = 5 x 3 + 5 x 4	**ESTIMATE** 1. To calculate approximately. 2. A number thought to be close to a number but possibly not exact. 3. To guess a value based upon knowledge.
EXPONENT 1. The number that represents the power. 2. It represents the number of times to multiply a number by itself. 3. In the number 4^3, 4 is the base and the small 3 is the ___.	**FACTORS** 1. Numbers multiplied together to get a product. 2. The ___ of 12 are 1, 2, 3, 4, 6, and 12. 3. The ___ of 18 are 1, 2, 3, 6, 9, and 18. The ___ of 24 are 1, 2, 3, 4, 6, 12, and 24. The GCF, or greatest common ___, of 18 and 24 is 6.
HUNDREDTHS PLACE 1. In the number 1,264.085 the numeral 8 is in this place. 2. In the number .6589 the numeral 5 is in this place. 3. If you round the number 869.169 to the ___, you get 869.17.	**IMPROPER FRACTION** 1. A fraction with a numerator that is greater than its denominator. 2. Examples are 86/9 and 5/3. 3. To change one to a mixed number, divide the numerator by the denominator. (The remainder will be the denominator of the fraction part of the mixed number.)

Number Sense Bingo

INTEGERS

1. Positive whole numbers, zero and the negative opposites of the whole numbers.
2. These positive and negative numbers can be shown on a number line on either side of zero.
3. Some examples are 0, 1, -1, 2, -2, 200, and -200.

INVERSE OPERATIONS

1. If one operation undoes the effect of another, they are said to be ___.
2. Multiplication and division are ___.
3. Addition and subtraction are ___.

IRRATIONAL NUMBER

1. A number that cannot be written as a simple fraction or as a finite or repeating decimal.
2. π is one because there is no pattern to the decimal and it cannot be written as a simple fraction.
3. The $\sqrt{2}$ is one but the $\sqrt{4}$ is not.

MIXED NUMBER

1. A number made up of a whole number and a fraction.
2. Examples are 9 5/9 and 1 2/3.
3. To change one to an improper fraction, multiply the whole number by the denominator and add the answer to the numerator.

MULTIPLE

1. The product of a number and any other number.
2. The LCM, or lowest common ___, of 2 and 6 is 6.
3. To find a common denominator of 2 or more fractions, you must find the lowest common ___.

NEGATIVE INTEGER

1. A positive integer x a negative integer = ___.
2. A positive integer ÷ a negative integer = ___.
3. A negative integer ÷ a positive integer = ___.

PERCENT

1. It means "out of 100."
2. If 30 people out of 100 prefer vanilla ice cream, we say that 30 ___ of the people prefer that flavor.
3. To change this kind of number to a decimal, remove the ___ sign and place a decimal point 2 digits to the left.

PLACE VALUE

1. Value given to a particular figure because of its position in the number.
2. Each digit in a number has a different ___.
3. The value of the position of a digit in a numeral is that digit's ___.

POSITIVE INTEGER

1. A positive integer x a positive integer = ___.
2. A positive integer ÷ a positive integer = ___.
3. A negative integer x a negative integer = ___.

PRIME FACTORIZATION

1. It is the breaking of a composite number into its prime factors.
2. The ___ of 100 is 5 x 5 x 2 x 2.
3. In exponential form, the ___ of 100 is $5^{2} \times 2^{2}$.

Number Sense Bingo

PRIME NUMBER
1. A whole number with only one set of factors: itself and 1.
2. It is divisible only by itself and 1.
3. Two is the only even number that is this.

PROPER FRACTION
1. A fraction whose denominator is greater than its numerator.
2. 3/4 is one; 4/3 is not.
3. 99/100 is one; 100/99 is not.

RATIONAL NUMBERS
1. Real numbers that can be expressed as a ratio, or quotient, of 2 non-zero integers.
2. Finite decimals, fractions, and repeated decimals are these.
3. π is not one.

RECIPROCAL
1. The number that is used to multiply by a given number to get a product of 1.
2. 4/5 is the ___ of 5/4.
3. 10/9 is the ___ of 9/10.

REDUCE
1. To find a smaller form of a fraction.
2. To do this, divide the fraction's numerator and denominator by the same number.
3. We do this when we change a fraction to an equivalent fraction in a smaller form.

ROUND
1. To shorten a number to a certain place value.
2. When we ___ 375.82 to the nearest tenth, we get 375.8.
3. When we ___ .569 to the nearest hundredth, we get .57.

THOUSANDS PLACE
1. In the number 98,221 the numeral 8 is in this place.
2. In the number 5,296.378 the numeral 5 is in this place.
3. The digit in this place is just to the left of the digit in the hundreds place.

THOUSANDTHS PLACE
1. In the number 6.2591 the numeral 9 is in this place.
2. In the number .987 the numeral 7 is in this place.
3. If you round .8542 to the ___ place, you get .854.

WHOLE NUMBERS
1. Positive numbers with no fraction or decimal point.
2. Positive integers.
3. The numbers 4, 21 and 16 are ___.
The numbers -4, 21 1/2 and 16.5 are not.

ZERO
1. Any number times this is zero.
2. The sum of any number plus this is that number.
3. This number is neither positive nor negative.

Number Sense Bingo

Improper Fraction	>	7/24	Estimate	Distributive Property
Commutative Property	<	Zero	6 2/3	Hundredths Place
1/2	Thousands Place		Percent	Inverse Operations
11/20	15%	Thousandths Place	36	Place Value
Positive Integer	400%	Decimal	Prime Number	Irrational Number

Number Sense Bingo

11/20	1/2	Exponent	Reciprocal	7
Place Value	6 2/3	4/5	15%	Multiple
1 3/4	400%		42	Thousandths Place
Round	Rational Numbers	Thousands Place	Proper Fraction	Irrational Number
Hundredths Place	Zero	Decimal	Commutative Property	Prime Number

Number Sense Bingo

11/20	Thousandths Place	6 2/3	36	1/2
400%	<	.875	>	Factors
15%	Zero		Multiple	-3
Thousands Place	1 3/4	Positive Integer	Round	Exponent
Prime Number	Commutative Property	Decimal	Proper Fraction	7

Number Sense Bingo

Thousands Place	Multiple	7/24	Commutative Property	Improper Fraction
Integers	4/7	>	Reciprocal	1/2
Percent	Round		Distributive Property	Estimate
Thousandths Place	48.307	Zero	Decimal	4/5
Associative Property	Hundredths Place	Whole Numbers	Prime Number	Inverse Operations

Number Sense Bingo: Card No. 4

Number Sense Bingo

Hundredths Place	Distributive Property	15%	4/5	Commutative Property
Integers	Thousandths Place	.875	42	<
7/24	Inverse Operations		Negative Integer	Absolute Value
Irrational Number	7	Improper Fraction	Proper Fraction	2 2/3
6 2/3	Decimal	1/2	Thousands Place	Percent

Number Sense Bingo: Card No. 5

Number Sense Bingo

-3	Multiple	Exponent	7	Inverse Operations
36	15%	2 2/3	>	1/2
Reciprocal	Associative Property		4/7	42
Decimal	Positive Integer	Proper Fraction	Whole Numbers	7/24
Place Value	Thousandths Place	Improper Fraction	Percent	48.307

Number Sense Bingo

Improper Fraction	Multiple	Absolute Value	Negative Integer	6 2/3
Place Value	7	400%	<	Integers
Exponent	Estimate		42	4/7
Thousands Place	Round	.875	11/20	1 3/4
Decimal	Commutative Property	Proper Fraction	Whole Numbers	-3

Number Sense Bingo

Percent	Multiple	9	36	4/7
Integers	7/24	Reciprocal	Inverse Operations	4/5
48.307	Prime Factorization		7	Distributive Property
Prime Number	Thousands Place	11/20	Associative Property	Round
Zero	Decimal	Whole Numbers	15%	Place Value

Number Sense Bingo

42	6 2/3	400%	48.307	Commutative Property
Associative Property	7	Percent	15%	Multiple
Factors	Improper Fraction		<	9
2 2/3	Irrational Number	Positive Integer	Negative Integer	Absolute Value
Round	Proper Fraction	.875	11/20	Distributive Property

Number Sense Bingo

11/20	36	4/7	Reciprocal	48.307
Inverse Operations	4/5	>	<	7
Prime Factorization	Multiple		Estimate	1 3/4
Positive Integer	Irrational Number	2 2/3	Proper Fraction	Factors
.875	Place Value	Exponent	Hundredths Place	Percent

Number Sense Bingo

-3	Multiple	15%	2 2/3	Place Value
9	Factors	Negative Integer	42	>
Integers	7		Exponent	400%
.875	1/2	Proper Fraction	Commutative Property	11/20
Associative Property	Decimal	Improper Fraction	Whole Numbers	6 2/3

Number Sense Bingo

6 2/3	Distributive Property	Factors	36	42
400%	Place Value	7/24	Whole Numbers	<
Improper Fraction	Absolute Value		Inverse Operations	Reciprocal
Decimal	Round	7	11/20	Integers
Multiple	9	Prime Factorization	Associative Property	4/5

Number Sense Bingo

2 2/3	Distributive Property	-3	Factors	Inverse Operations
7/24	9	7	42	1 3/4
36	4/5		400%	Absolute Value
Percent	Proper Fraction	4/7	Prime Factorization	11/20
Decimal	Irrational Number	Whole Numbers	Improper Fraction	Negative Integer

Number Sense Bingo

Commutative Property	7	15%	42	Associative Property
4/5	Improper Fraction	Factors	<	Multiple
2 2/3	Estimate		Exponent	.875
Irrational Number	Proper Fraction	Prime Factorization	4/7	-3
Decimal	Reciprocal	1 3/4	Place Value	Percent

Number Sense Bingo

Negative Integer	42	15%	6 2/3	36
-3	Exponent	>	7/24	Associative Property
Inverse Operations	Improper Fraction		1/2	Multiple
Decimal	Factors	9	Proper Fraction	2 2/3
Place Value	Round	Whole Numbers	48.307	400%

Number Sense Bingo

4/7	Factors	9	48.307	Rational Numbers
Reciprocal	1 3/4	Absolute Value	Integers	Estimate
2 2/3	Distributive Property		Inverse Operations	400%
Thousands Place	4/5	Decimal	Negative Integer	11/20
Associative Property	Reduce	Whole Numbers	Round	Multiple

Number Sense Bingo: Card No. 16

Number Sense Bingo

.875	Mixed Number	Composite Number	Factors	Commutative Property
Negative Integer	Associative Property	Proper Fraction	Estimate	Absolute Value
42	Percent		Reduce	9
Irrational Number	Place Value	11/20	15%	1 3/4
Positive Integer	2 2/3	6 2/3	36	Distributive Property

Number Sense Bingo: Card No. 17

Number Sense Bingo

48.307	Prime Factorization	4/5	2 2/3	Reciprocal
Multiple	.875	Positive Integer	Inverse Operations	Associative Property
42	1 3/4		Composite Number	7/24
Irrational Number	>	Proper Fraction	11/20	Exponent
Reduce	Factors	15%	Mixed Number	-3

Number Sense Bingo

Inverse Operations	-3	Factors	9	Prime Factorization
Negative Integer	36	Multiple	6 2/3	Estimate
Mixed Number	Commutative Property		<	1/2
Exponent	Reduce	Positive Integer	Round	Composite Number
7/24	Rational Numbers	Place Value	Percent	Whole Numbers

Number Sense Bingo

Prime Factorization	Mixed Number	36	Factors	Whole Numbers
4/5	400%	Integers	Positive Integer	Reciprocal
Distributive Property	Absolute Value		Thousands Place	>
Hundredths Place	Zero	Prime Number	Round	Reduce
Thousandths Place	Percent	Rational Numbers	11/20	Composite Number

Number Sense Bingo: Card No. 20

Number Sense Bingo

Negative Integer	-3	Integers	Factors	Hundredths Place
Distributive Property	Composite Number	4/7	9	Absolute Value
1 3/4	Place Value		Mixed Number	15%
Positive Integer	6 2/3	Reduce	Irrational Number	Percent
Thousands Place	Rational Numbers	Whole Numbers	.875	Round

Number Sense Bingo

48.307	Exponent	Composite Number	7/24	2 2/3
Reciprocal	36	1/2	9	<
4/5	Estimate		Improper Fraction	Absolute Value
Reduce	Irrational Number	Round	>	Commutative Property
Rational Numbers	.875	Mixed Number	1 3/4	Integers

Number Sense Bingo

4/7	Mixed Number	6 2/3	7/24	Whole Numbers
-3	Prime Factorization	Place Value	Negative Integer	>
Exponent	2 2/3		Prime Number	Improper Fraction
1 3/4	Rational Numbers	Reduce	.875	Round
Hundredths Place	Zero	Percent	Positive Integer	Composite Number

Number Sense Bingo

4/7	Prime Factorization	Commutative Property	Mixed Number	9
Composite Number	Whole Numbers	Integers	Reciprocal	Improper Fraction
Absolute Value	48.307		2 2/3	1 3/4
Hundredths Place	Prime Number	Reduce	.875	Distributive Property
Thousandths Place	Thousands Place	Rational Numbers	36	Zero

Number Sense Bingo

Thousands Place	Integers	Mixed Number	15%	Composite Number
>	Irrational Number	Negative Integer	4/7	<
Distributive Property	9		Prime Number	Reduce
1/2	Hundredths Place	Zero	Rational Numbers	Estimate
Whole Numbers	Commutative Property	4/5	Associative Property	Thousandths Place

Number Sense Bingo

Composite Number	Mixed Number	Exponent	Reciprocal	48.307
Positive Integer	36	9	Prime Factorization	4/7
Irrational Number	Prime Number		Estimate	Thousands Place
.875	7/24	Hundredths Place	Rational Numbers	Reduce
Absolute Value	Associative Property	15%	Zero	Thousandths Place

Number Sense Bingo

Exponent	4/5	Mixed Number	Prime Factorization	400%
Hundredths Place	Prime Number	Negative Integer	Reduce	<
Proper Fraction	Zero		Rational Numbers	Thousands Place
48.307	-3	Integers	Thousandths Place	>
Associative Property	Estimate	Composite Number	1/2	Absolute Value

Number Sense Bingo

Inverse Operations	Prime Factorization	1/2	Mixed Number	4/7
400%	Composite Number	Prime Number	Reciprocal	Estimate
Zero	1 3/4		Absolute Value	Positive Integer
11/20	48.307	Place Value	Rational Numbers	Reduce
7/24	42	Associative Property	Thousandths Place	Hundredths Place

Number Sense Bingo

Composite Number	Prime Factorization	48.307	Negative Integer	42
Irrational Number	Positive Integer	Integers	Absolute Value	1/2
Distributive Property	Prime Number		<	Mixed Number
400%	Hundredths Place	7	Rational Numbers	Reduce
4/7	9	Thousandths Place	-3	Zero

Number Sense Bingo: Card No. 29

Number Sense Bingo

Commutative Property	Mixed Number	Reciprocal	42	Reduce
>	Prime Factorization	Exponent	Estimate	<
Irrational Number	2 2/3		Absolute Value	Integers
Thousandths Place	-3	7/24	Rational Numbers	Prime Number
Hundredths Place	6 2/3	Zero	Composite Number	1/2

Number Sense Bingo: Card No. 30

www.ingramcontent.com/pod-product-compliance
Lightning Source LLC
Chambersburg PA
CBHW051419200326
41520CB00023B/7299